Yes, But Do They Understand? How Children View The World

Topic #1: Trees Are Made Of Air

Cultivating Children's Reasoning

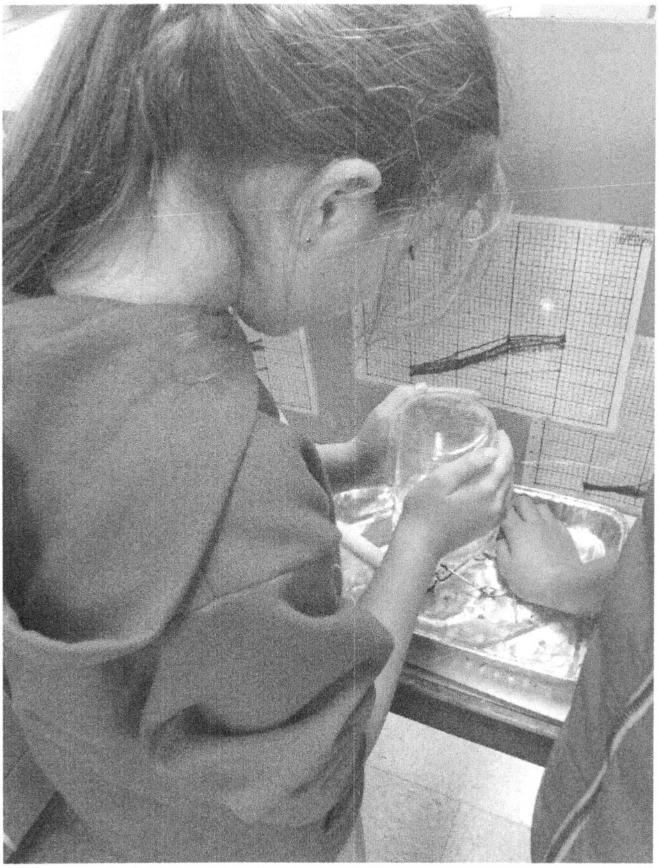

MEREDITH OLSON, Ph.D.

Trees Are Made Of Air

www.DocOsBooks.com

Printed in USA

Published by:
SAN 299-2701
Glenhaven Publishing
4262 NE 125 Street
Seattle, Washington 98125

ISBN 978-0-9984627-5-2 (paperback)

Table Of Contents

Trees Are Made Of Air

But First ...

We are going to talk about candles.
But, wait, isn't this book about trees?
It is, but we need to start at the beginning. Therefore, we will start with candles.

You, the reader, know our goal is to talk about trees but I do not tell my students. I just start having them explore candles. I intentionally set them up so that it is possible for at least a few of them to make connections later.

There is a method to my madness. A method to the lesson style. If I just tell them, they won't remember it. Do you remember? What are trees made of? How does that work? Why don't you remember? You were taught about it over and over for twelve years. Perhaps you learned the specifics – but somehow the "broad brush" overview escaped understanding.

We all forget specifics. But remembering general concepts is what takes us through life. My goal is to implant overarching images that last into adulthood.

Here we explore the awareness of children. How they see the world. We will try to understand how they resist changing their immature conceptualizations. These pages give examples of how I "set children up" so they can make connections between ideas. The goal is to produce people who feel competent and capable. Who feel they can (sometimes) connect ideas and carry with them an overarching framework into which they can place future learning.

And so we start with candles. A major goal is to present the ideas of *physical change* and *chemical change*. Candles exhibit both types of change. That catches children by surprise. They can't imagine that things can both burn and melt. It must be one or the other. Not both. What a surprise!

The real issue is how to go about understanding another person's beliefs. I give "half page" tests. I stand in front of the class and cut paper in half. That creates a mindset: "*Oh no, she is going to ask us something*!" Half-page tests get them to engage in the topic. It creates awareness of what they each believe at that moment. By avoiding class discussion, each child is free to try to dredge up what they actually believe. Children are suggestible. If I allow prior verbal discussion, they will quickly all agree. If, when asked why burning things turn black, and some child says out loud the word "soot", all of the children will incorporate soot into their explanations. I have to require the room to be silent. No prompts. So I capture their deep misconceptions. Ideas they don't even know they hold. I repeatedly say, *"Do your best. It is OK to be wrong." It isn't useful to say, "I don't know." That means you haven't memorized the idea. It is better to say, "I never thought about that, so I don't know – yet." And then to add, "But maybe…".*

Science is a subject where it is OK to be wrong. Science can never prove we are correct. Confidence intervals and t-tests tell us the likelihood that we are wrong. When we are wrong, we are fairly certain we are wrong. Trying to be correct is always a theory. The theory seems to work for now. So, wrong answers are intelligent responses in science. The students need to feel psychologically safe enough to pose thoughtful, wild, possibly incorrect answers. Being wrong leads to the possibility of having a new idea. It leads to the belief that, *" I have the capacity to have an original idea."*

So, now candles. How can I cause kids to notice that burning wax goes into the air. How can they think about air? Will it become possible for them to think carefully about air and conclude what air has to do with growing trees? This is an exploration. An adventure into the craft of teaching your children. Can I do it? Or will I just have to end up telling them? We will see.

Candles

My students just completed a month long study of the Gut which gave them a self image of competence and intelligence. Of being able to memorize and understand an enormous amount of medical terminology. Being able to draw, from memory, a magnificent medical drawing with fifty to a hundred labeled parts. Although we did a lot of dissections of parts of the gut, (tongue, larynx, trachea, esophagus, stomach, lungs) much of the time involved transmission, reception, absorption and rehearsal of terminology, structure and function. Because I refuse to give homework, and we use no textbooks, this was accomplished within a 43 minute daily class period of discussion and lab activity for about a month. They did a great job and parents were proud to take home each child's drawing at conference time.

I have found that children benefit from experiencing different types of lessons that require different types of learning. Memory. Reasoning. Data analysis. This creates variety and gives them opportunities to learn how they learn. We just had a productive month of expository learning. Now it is time for another type of learning. Inquiry. It might be called interactive, constructivist learning. A process of learner centered conceptual construction with the teacher continually probing for understanding and misconceptions.

Trees Are Made Of Air

Michael Faraday

Michael Faraday was an important person in the history of science. He studied electromagnetism and electrochemistry and invented devices which began our electric motor technology. He was an excellent experimentalist and a popular lecturer. He gave Christmas lectures for young people starting in 1825. The Chemical History of a Candle was the title of six lectures Faraday gave at the Royal Institute in London in 1848.
https://en.wikipedia.org/wiki/The_Chemical_History_of_a_Candle

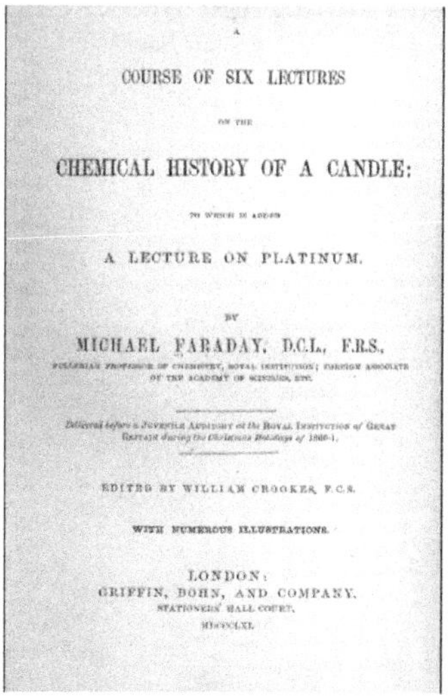
A
COURSE OF SIX LECTURES
ON THE
CHEMICAL HISTORY OF A CANDLE:
[illegible]
A LECTURE ON PLATINUM.
BY
MICHAEL FARADAY, D.C.L., F.R.S.,
[illegible]
[illegible]
EDITED BY WILLIAM CROOKES, F.C.S.
WITH NUMEROUS ILLUSTRATIONS.
LONDON:
GRIFFIN, BOHN, AND COMPANY,
STATIONERS' HALL COURT.
[illegible]

Thinking about Faraday gives status to our study of a candle. Students were eager to learn.

But it is not very engaging to just look at a candle and try to notice things.
A child needs a task
OK. Who can burn candles fastest?

I want them to expose their belief about candles. I have noticed that children of this age think candles don't burn. They just melt. In past classes, they almost uniformly believed that the wick burns. They believed wax could either burn or melt. But not both.

Probably wax just melts. When they heard me say the word burn, their minds accept the meaning "melt". How can I confront them with their misconception?

Having students use flame in the classroom can be dangerous. I have to be very clear, very specific, about what they are allowed to do. They each write the lab in their journals while I write it on the overhead projector. Problem, Equipment, Procedure, Results. Each lab station will gather materials such as a foil cake pan, alcohol burner, match petri, candle, and fire-extinguisher beaker from the supply counter.

Calibrate a Candle

"Burning fastest" implies a race. How much will burn in a minute? We have to calibrate a candle in centimeters so we know how long it is at the start and how long it is at each 30-second period. Besides, scratching marks in the wax gives practice in using centimeter measurements. It also draws a student's attention to the idea of precision in measuring.

When the materials are assembled, they call out: "Ready to light, Doc 'O'." No one lights a match without me standing next to them so I can grab their wrist and turn their hand, if needed, so the flame is higher than their hand.

Each lab, eight per class period, light a candle and let it burn for five or six minutes. I admonished them to be sure to hold the candle over the foil pan in case any dripped. Wax drips on the floor make the room slippery. My saying this also implied permission to let some wax drip. I wanted to allow their misconceptions to reveal themselves.

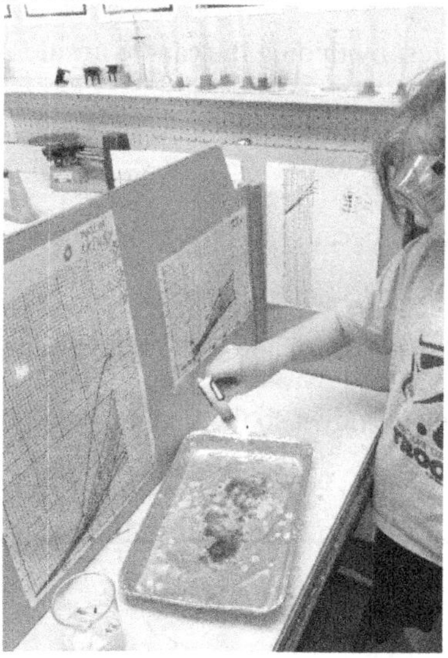

I even went so far as to ask them to write down their strategy for burning a candle fastest. Before doing the experiment, on a half-page test I asked, **"How would you burn a candle fastest?"** The students all wrote the question on their paper so I knew they had heard the term "burn". Many then proceeded to use the word "melt" in their response. This allowed me to ascertain whether students used the terms "burn" and "melt" interchangeably or whether they were aware of the difference. As I expected, students were not yet aware of the idea of burning. Most of them did not make a differentiation between the concepts of burning and melting.

Student ideas: How Would You Burn A Candle Fastest?

Light the candle on both sides of the candle.

I will hold the candle so gravity will help.

Hold the candle upside down with a test tube holder. My reasoning is that the flame will curl around the candle, so all the molecules will bump into each other faster.

I would burn a candle fastest by tipping the candle upside down and making sure it drips so wax melts quicker.

I would hold the candle upside down.

My plan is to face downward because heat rises and the flame might rise faster too.

I would hold the candle upside down and it should melt faster.'

I would turn the candle upside down so the flame would point at the wax and melt it.

Trees Are Made Of Air

To burn the candle fastest you have to flip it upside down.

I would turn it upside down.

I will burn the candle by flipping it upside down so so that the flame melts the wax faster.

I would hold the candle and light, possibly rotate.

I would turn the candle over and blow really lightly on it.

I would turn the candle upside down so it would burn the wax faster.

Blow hot air on it?

I would put the match to the side of the candle and the wick just to speed up the process and to be faster.

I would turn the candle upside down so the flame would burn the wick and the wax so it would go faster.

Heat up a skewer with matches and rest the candle over the skewer while the wick is burning. TaDa!

Light the candle with the matches from the match petri.

Wrap the candle with foil.

I would hold it at an angle so the wax gets more runny and drips faster, so that the candle gets stubby faster.

Light it and blow on it and hope.

I would tilt it to the side so the flame would be closer to the candle melting it faster.

I would burn a candle by lighting it, holding it down and waiting for it to get shorter and shorter. I would hold it down because then the flame would travel upwards faster. At least that's what a match does.

I would put Tyvek against the candle so it covered it so it would heat the wax faster.

To burn a candle fastest I would light the candle and wait for the wax to melt all the way.

I would blow on it.

I would burn both ends.

I can't think of anything. I didn't know you could burn a candle.

I think you light the candle because candles melt, they don't burn. Only the wick burns.

What if you put the match to the wax and melt it?

Light something else below. The hotter the better.

I will try to hold the candle upright so all the wax will melt evenly.

Light the wick lower.

Is it how much wax drips down?

Blow on it very lightly so the Oxygen feeds it.

Take the candle and turn it upside down with a test tube holder.

Put a glass slide close to it.

Turn the candle upside down.

I would hold the candle upside down.

Light a match and them light the candle.

The next day they scurried into class ready to report findings from their exciting lab. After bringing their "evidence" (pan and candle) to their writing desks, each lab reported the ending candle length and the amount of wax collected in the pan. Each student recorded the findings from every lab in their notes. Reflecting on the data and arriving at conclusions, students proudly noted that they had shorter candle than others. Some had a lot of wax in their pan – and much of the wax was black.

Aha. An opportunity. The candles we used were 12cm standard white camping candles. Where did the blackness come from?

I didn't dare ask them the question out loud because students of this age are very suggestible. In my experience, once a student says "soot", everyone will say that. So, I just pass out half sheets of paper in silence. **"Oh no, another half-page test!"**

Typically tests are given to generate grades. I never do that. I don't use tests to judge. I want to know how to teach. I need to figure out where they are. What I do next is based on their awareness so I can help them through their process of understanding. In order to do this I need to learn where everyone thinks the blackness came from.

I ask, **How can a white candle produce black wax?**

Student ideas: What makes white candles turn black?

When you burn firewood, it chars and burns to ashes. I think the candle burnt, and charred because its material and influence by some other wax and the flame turns things black.

Because fire burns and chars things because it makes all the moisture in it evaporate and they dry up.

Fire chars the wick.
Fire chars the wax.

I think it is because when you light a candle the fire lights the wick and burns the pieces of string together and gives it a charred look. I think that char gets into the wax and because the wax is white, it turns it gray. I don't know how.

Both the wax and the wick may have some interaction with Oxygen or heat.

If any of the wax is touched by the flame, it becomes dark, if it is held by a flame for a long time, then it becomes blacker and blacker. When the flame burns the wick, because of the heat, it loses its chemical properties and turns black.

The molecules in the wick are heat sensitive but the wax molecules are less heat sensitive.

Because smoke from the fire went into the candle making it black. The same thing with wax.

The wax got burnt. The wick got burnt.

The burned black wick fell off and mixed with the white wax.

Maybe when the wax gets hot enough, it turns black. It burns. The flame burnt the wick.

I think the wax burned like wood because when wood burns it gets black with charcoal. The wick burned just like the wax.

Maybe the angle (you hold it at) changes how hot the wax gets, so it turned black. The wick burnt.

Trees Are Made Of Air

When something burns it turns black, like fire wood. It starts out brown but it is black charcoal after burning.

Maybe the fire burnt the wax and the wick.

The wick got black because the wick was originally dipped in wax, then when the wax melted, the wick burned and the new black wick dropped onto the wax and turned the wax black.

When you hold the candle upside down, the smoke probably makes the wax black and it melts into liquid and black wax drips.

Charring of the flame turned it black.

It melted and exposed to cold air. Burned by the flame and charred.

The black wick was burnt by the fire and they held it upside-down so the char went up in the flame and turned the wax black.

Because when the fire burns the energy wears out so it turns black.

It ended up like that from ash mixed with the wick. The wick is just string covered in wax so when all the wax burned off it got black.

Maybe because it burned for awhile an some of the wax dripped off but the wax and wick were just burnt.

The fire burning the wick makes it black then it melts the wax and turns the was black.

Fire burning the wax. Wire burning the wick.

The flame burned the wax and the wick so they turned black with the jumpingness (of heat).

The wick gets burnt so it turns black and when it is on fire it tilts turning the wax black.

When fire burns wood it turns black, but it turns gray first. Maybe it turns things it burns black because it burns Oxygen.

Trees Are Made Of Air

I think it is from the wick burning and becoming black and then becoming ash which goes on the wax and makes some of the wax black.

When the wax burns the flame chars the wax and burns.

Heat the candle, or just the heat. Because I think when something burns, it turns black (mostly) also maybe its because of the candle's structure.

The blackness comes from the wick.

It comes from the fire burning the wax until it was grey or black, and the hotter the candle the darker the wax.

When we lit the wick of the candle, it turned black. Then the candle got smaller. The wick goes through the candle, turning the candle black and turning the wax drops black.

The wax burns and then drips gray wax because the wax is burning.

I think it was from smoke from the fire which some smoke fell in the tray and some smoke fell into the candle.

The smoke dyed the wax brown/grey and black.

Smoke produced by the fire.

I think there is black wax because of the ash. When something burns it turns black because of that. The wax dribbles down the candle and when it dries it becomes black because of the ash.

Maybe the heat discolored the wax. Maybe the candle charred.

I think the blackness came from the metal on the test tube holder, because once we get to the holder it turns black and gray.

Blackness came from the test tube holder's metal.

Black Wax From A White Candle

But then I said, *"Lets look at the data again."*
"You lit the candle for five minutes."
"Many labs reported wax drops in the pan."
"What?"
"Wax drops in the pan?"
"Oh No" "You were told to burn a candle." How did you get wax in the pan?"
"Candles melt."
"Candles don't burn. They melt." "Just the wick burns."
"But you were told to burn a candle."

Now things were getting interesting.
Suddenly the students felt they had done something wrong.
But what? Were they tricked by their teacher? … Again?

We were asked to burn a candle!
What is burning?
Not owie. Not *"ow", "I burned my finger."*
What kind of things burn?
Wood burns. Campfires. Forest fires.
Clothes burn.
Houses burn.
But rocks don't burn.
Students were puzzled. No one was sure what burning really is.

Is it possible to burn a candle without getting any wax drips? Students tried the lab again but this time were careful to hold the candle upright. They used a skewer to scoot any potential drips back to the top of the candle. Most students noticed a little bowl shape formed at the top of the candle. Several noticed the wick bent over in an upside-down j-shape. One student commented that the base of the wick remained white. Just the upper part turned black. Students burned their candle for five minutes.

The next day each lab reports their finding. All the candles but one have become a few millimeters shorter. One candle had a long wiggly wick that stuck out the bottom of the flame. Their candle failed to make a bowl shape and wax insisted on dripping from the lopsided top.

What Is Burning?

Lets think about a Yule log.

The Yule Log was originally an entire tree, that was carefully chosen and brought into the house with great ceremony. The largest end of the log would be placed into the fire hearth while the rest of the tree stuck out into the room! The log would be lit from the remains of the previous year's log which had been carefully stored away and slowly fed into the fire through the Twelve Days of Christmas.

https://www.whychristmas.com/customs/yulelog.shtml

https://bjws.blogspot.com/2016/12/the-winter-solstice-yule-log.html

After burning for twelve days, only ashes were left on the hearth.
Where did the gigantic log go?
Smoke?
OK. What is smoke?
Why is smoke black?
If I burn your shirt, it becomes black.
If I burn a white towel, it becomes black.
Why black?
Why not pink? Or some other color?
Remember when we were studying the Gut? We chewed a soda cracker. What is a soda cracker? We used the Iodine test to show that it is starch – and it can be changed to sugars.
Carbohydrate.
We call that "*worms-with-clothes-on*".

```
  H H H H H H H H H H H H H H H H
H-C-C-C-C-C-C-C-C-C-C-C-C-C-C-C-C-H
  H H H H H H H H H H H H H H H H
```

Trees Are Made Of Air

Will a soda cracker burn?
What color does it become when it burns?
Well, this is confusing. It turns black from Iodine and from burning – but the two are different processes.

Lets forget about Iodine for now. Burning a soda cracker makes it black.
Burning a carbohydrate makes it black.
Where do carbohydrates come from?
Plants.
Burning a plant makes it black.
But why?
Is blackness painted on the outside? Many students think so.

It is time to tell them something. They are ready.
The best explanation I know of is that heat of the fire "rips their clothes off".

Heat is "jumpingness". Students know this from previous lessons. The heat of a fire makes the carbohydrate molecules jump so violently that they jump right out of their Hydrogen clothes.

```
  H H H H H H H H H H H H H H H H
H-C-C-C-C-C-C-C-C-C-C-C-C-C-C-C-C-H
  H H H H H H H H H H H H H H H H
```

What is left?
Carbon!
Naked Carbon bodies.
Burned things are black because they reveal their black naked Carbon bodies!

Naked bodies. Without Hydrogen clothes.
That's where the blackness comes from.

Sweaters turn black.
Houses turn black.
Forests turn black
All carbohydrates are made from C&H.
"Worms-with-clothes-on."
Rip off the clothes and you can see the naked black Carbon.

We see smoke. That is Carbon. Can we catch it so we can collect evidence? Can we catch the "naked bodies" jumping around?
Well, that is progress.
We have a visual image of what burning is.
But, back to the question: "Do candles burn?"

We got some black wax in our pan from the first lab we did.
Maybe something was burning.
Was it the wick?
Was it the wax?
How can we design a lab to tell?

Trees Are Made Of Air

Do Wicks Burn?

Get a foot long piece of string. That is a lot like a candle wick.
Will it burn? It is made of cotton so I think so.
Cotton is a plant and plants burn.
We write up a lab. Hold the doubled string in a test tub holder over the cake pan and ignite the lower end with a match. Hold a flask by the neck and hold the bulb in the rising smoke. Can you catch "naked bodies" on your flask? Of course, you must have the teacher standing at your side as you do the lab.

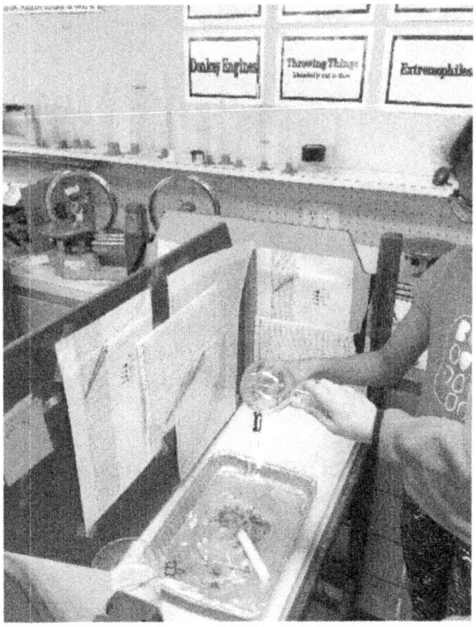

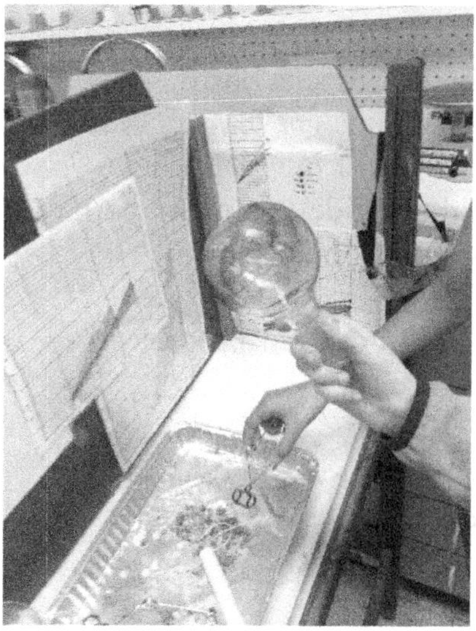

Oh my! Catching "naked bodies" in science class!
YES, YES! That worked. Each lab tried it. Students were able to catch small spots of blackness on their flask. All but one successfully caught some "naked bodies". They left the room saying, "That was the coolest class!"

Students gathered in class the next day with their soot-stained flasks. They estimated how large a spot of blackness they had grabbed. Each lab reported their finding and held up their flask. Yes, we have evidence that we can get "naked bodies" from string. String burns. Wicks must burn.

Where did all that blackness come from?

Lets review what we have found out.
Wicks are plants. "Worms with clothes on". Carbohydrates.
When they burn they leave black "naked bodies". Carbon.
So, what is our evidence?

Do you believe wicks burn?
"Yes, because we get blackness on the flask."
"But wax doesn't burn? No, wax doesn't burn."
What kind of evidence would you need to convince you that wax does or does not burn?

I need to monitor their progress. Are they modifying their concepts?

Another half-page test! What do they now think about wax? **Does wax burn?**

Student ideas: Does Wax Burn?

The wax must burn because the candle got shorter and there isn't any melted wax.

No, because when we burnt the candle we put the flask in the flame, and when we burnt the wick we didn't put the flask in the flame.

I don't think wax burns. I think that it actually traps the naked bodies (Carbon) until you touch it to the flame.

Wicks burn because when we held the flask above the burning wick it did grab naked (Carbon) bodies.

Wax burns because when we scooted the wax back up the candle it disappeared.

I think candle wax melts, not burns. The wick burns a little bit after the wax burns.

Wax burns but not very well. It is there to slow down the burning of the wick so it lasts longer.

I think some of the wax burns because the wax piles up in the middle, but the candle gets shorter too.

Candle wax doesn't burn but it slows down the wick as you get more naked Carbon) bodies.

Wicks do not burn because they don't disappear. Candles do burn because if you leave them for long enough and don't let wax drip, the wax will disappear.

The wick burns and the candle melts.

The wick burns because it turns black.

Wicks burn. They get shorter. Wax burns. Candles get shorter.

Wax burns and the naked bodies are very small. The wick burns because when it is blown out the wick isn't longer than the wax.

The wick burns because it disappears. The wax burns because it disappears.

I think when the wick is burning and there is a wick the candle only melts.

The alcohol on the wax on the string burns. Wax doesn't burn, it melts.

The wax preserves the wick but it still burns the dame way as just the string, but slower, because of the wax. All that wax that is melting goes down the wick, although we can't see it burning.

A wick burns because it turns black and that means that clothes are being ripped off.

I don't think wax burns because I think it melts but it also depends if there are any carbohydrates in wax.

If the wax turns black then the wax just burned, but if it stays clear white then no it didn't burn.

Wicks burn. Wax doesn't.

Wicks burn.

I think a little bit of the wick burns but not all of it. I don't think wax burns.

Wicks burn. Not wax.

Wicks do not burn. Wax does not burn.

The wick doesn't get shorter when it is burning. It just gets black. Wax doesn't burn. It just melts from the heat.

I think the wick is not burning. I think the wax burns because it makes a bowl and produces smoke.

I don't think the wick is burning. I think it is turning black and twisting from the het. I think the wax is either burning or melting during the time the candle is on fire.

The wick is burning because we caught naked (Carbon) bodies. The wax is not burning because it can't catch fire and it is still making a puddle of wax.

Trees Are Made Of Air

Yes the wick is burning but the reason it doesn't burn as quickly as a long wick is because of the wax that slows it down. The wax is burning as well bur some burns and some melts.

I think the wick is burning because when it is on fire it turns black.

I think the wick is burning some because it becomes shorter but it is burning less than the wax. We get more naked (Carbon) bodies when we burn wax.

The wick burns because it is part of the candle. The wax burns in a way that is not like when you burn the wick.

The wick burns and the wax protects it from burning. Wax melts and burns. It is hard to explain but when toe wax melts the wick can't burn.

The tip of the wick is burning but the rest of the wick os only just hot. Some was burns and other wax does not heat up fast enough to burn and it melts.

I think the wick is burning because it disappears.

The wick burns because it is disappearing. The wax burns because the flask has more dead (Carbon) bodies than fro just the string.

Yes the wick and the wax burn.

I think the wick and the wax burn.

Trees Are Made Of Air

Actually, Wicks Don't Burn Much

In previous years most of the students had gone on a field trip to Pioneer Farms where they were able to use tools from days gone by. I asked them if they remembered the scissors kept near the candle holder used for trimming candle wicks after each burning. If wicks burned, why did they need to be trimmed after each burning? Asking for written response on what three things do flames need to keep going, every students was able to say; fuel, air, and heat. So, what was the fuel? The wick? How could such a small thing as a wick keep the flame going more than ten minutes.

Then we examined the shape of the wick. I told them that early wicks just stuck up above the flame and got in the way. The wicks we saw curved over at right angles after going up about one centimeter. When flame existed, we noted the very tip of the wick had red glowing embers. Amazing! Automatic scissors! What a clever invention to trim the wick as the candle burned!

But is the wick burning? How can we know if only the tip of the wick is burning? We held a toothpick in the top of the flame and found a singed spot on the toothpick where the flame had been. Then each lab held a toothpick low in the flame. We discovered that there were now two singed spots with an unburned section of toothpick in the middle. Interesting!

https://en.wikipedia.org/wiki/The_Chemical_History_of_a_Candle#/media/File:Candleburning.jpg

That must mean the flame only "burned" on the outside of the flame. The interior was not actually burning! The part of the flame touching the wick was not burning! How to account for this? Well, if the flame was all around, and if flame needed Oxygen, maybe no Oxygen could get to the center of the flame. Flame needs three things: fuel, air, heat, but the center of the flame had only two.

So, if the wick only burns at the tip, what is the fuel for the flame?

Does Wax Burn?

We could try to catch "naked bodies" again but this time hold the flask over a burning candle. If we get more blackness than we got from the wick, what would that mean?

Lets write up the experiment and try it.
Ok. Hold a flask over a candle flame. Touching it.
Turn the flask until it is entirely covered with velvety blackness.
Oh, my goodness! We get many flasks completely covered in velvety black flasks. There is lots more blackness. Lots!

Twenty- three flasks were completely covered in marvelously velvety blackness.

Is this evidence that wax burns?
"*No! You are tricking us!*"
"*There are uncontrolled variables. We held the flask in the flame longer.*
We got more blackness from the longer time."

I expected the students to move easily to the conclusion that more blackness would mean that wax actually burns. Not so.

Hummm. How to proced?

In each class, many students protested that the wax somehow insulates the wick so it burns longer and produces the extra blackness. They tenaciously held on to their prior conceptions.

It is temping to just tell them. To have them memorize the fact that wax burns. But, I am intrigued by what they are thinking. Why do they resist the conclusion? What is missing in their chain of logic?

The volume of smoke doesn't impress them.
Although they can work math lessons successfully, they don't readily apply amounts to the smoke they see.

Lets think about the black smoke.
What happens to the smoke from an old fashioned coal burning train?

https://ayearningforpublius.wordpress.com/2013/01/01/let-me-off-the-stupid-train/

One chimney spews black smoke, the other steam from the boiler. What happens to the smoke? Does it only become thinner and thinner? Every one of my forty-six current students thought so.

Is there anything else it can do to disappear? When it was wood it didn't look black. Why? Because it had Hydrogen clothes on. Can the black smoke put on clothes? What is in the air that could be used as clothes? The first two periods had no idea. One student in the third period suggested Oxygen.

Oxygen. Would that work? Well maybe it might make Carbon Dioxide. But, my students said, that would be bad. TV news tells them it is bad. Isn't Carbon Dioxide a poison? Isn't it bad for you?

Hummm. I said, *"But you breathe out Carbon Dioxide every minute. Does that make you all sinfull?"* There was shocked silence. I could see them struggling internally. *"No, I am not a bad person because I breathe. Producing Carbon Dioxide is not a sin. It is OK for me to breathe. I have a right to breathe."*

Actually, there is Oxygen in the air. Some of the Carbon in the smoke puts on Oxygen clothes and becomes invisible CO2.

Lets pay more attention to how a candle actually burns.
How is it useful that wax melts and burns?
Solid wax doesn't burn very readily.
Liquid wax doesn't burn very readily.
Wax vapor is what does most of the burning.

So now look very closely at a candle.
You will see that the wick ignites and burns down until wax melts and forms a little liquid puddle.
Then the liquid wax climbs the wick and evaporates. Remember adhesion. Remember making "laundry"? Paper chromatography? We spent a month last fall doing labs focused on Adhesion and Cohesion. Well, the same thing happens here. Liquid climbs the wick and evaporates. The hot tip of the wick ignites the wax vapor so the vapor burns.

If I hold a flask just in the edge of the flame I can catch Carbon bodies before they "put their clothes on" (their Oxygen clothes) and turn into invisible Carbon Dioxide.

That is amazing!
Just watch a candle do that!

And if I tip a candle sideways so the puddle drips down, not much liquid wax goes up the wick and the flame usually goes out.
"I get it!"

I decided to wait a while and see what else students could understand. I knew the final understanding will come from grasping the equation for photosynthesis.

$$CO2 + H2O \xrightarrow{\text{sunlight}} C\&H + O2$$

We went to work on another part of the equation.
Instead of thinking about Carbon, lets think about water.

Did Anyone Notice Water?

Did anyone notice water forming on the flask?
Has anyone ever watched a forest fire? There is smoke above it but there are also clouds. Water.

How can water collect over a candle fire?
Hold a 600ml beaker upside down over a candle flame.
What happens?
We get water vapor in the jar.
I thought water put out flames. How can we generate water from a flame? Yet it seemed to be happening.

The next day, try to collect water over an alcohol burner flame.
The 600 ml beaker filled with steam rapidly. Alcohol burners seem to give off even more water than candles.

Trees Are Made Of Air

So, what is in the alcohol burner? How does an alcohol burner work? How can it be that we have used the same wicks in the little burners for ten years? Why don't the wicks burn up?

Well, actually, an alcohol burner works a lot like a candle. The fuel soaks up the wick (by adhesion) just like the melted wax does. Above the wick the burner fuel evaporates. It is the vapor that burns. The wick only burns if you run out of fuel.

But that leads to the question: what is burner fuel? It is Ethanol. A chemistry book shows the diagram of two Carbon atoms surrounded by six Hydrogen atoms. Aha! Ethanol contains Carbon! And it gives off water when it burns. That's what the equation says!

Alcohol burner fuel is a hydrocarbon. Just Hydrogen and Carbon. Jet airplanes use a hydrocarbon fuel. Kerosene.

Think of jet contrails. What are they? They are water clouds and ice. Actually the contrails contain more volume of water than the volume of fuel burned. Burning a gallon of jet fuel (Kerosene) makes more than a gallon of water. One ton of fuel creates 1.24 tons of water in the exhaust. One pilot reports his A330 jet creates 7 tons of water per hour during cruise and about 13 tons per hour on takeoff.

Here is my KC-10 with three contrails coming from the three engines, forming some distance behind the exhausts

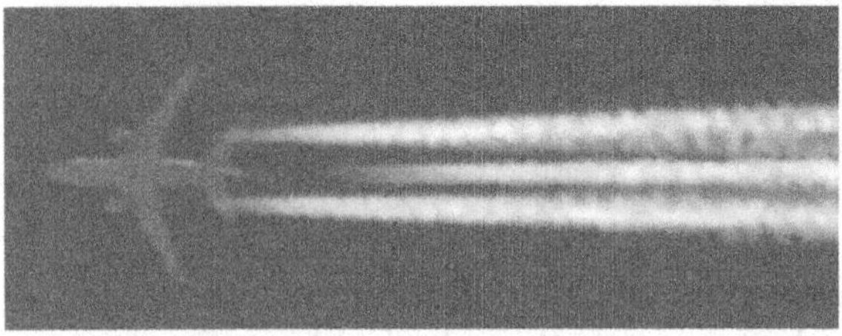

https://www.metabunk.org/how-much-water-is-there-in-jet-engine-exhaust-about-1-3-gallons-per-gallon-of-fuel-used.t4018/

Well, lets think about that.
We need Oxygen to burn things. C and H things.
I suppose if Carbon and Hydrogen are jumping around in the air and Oxygen is jumping around too, then some Hydrogen can hook-up with Oxygen. What is the formula for water? Hydrogen and Oxygen. Yes, we have the components to make water. $H2O$
And we can make Carbon Dioxide too. Carbon and Oxygen. $CO2$

A fire gives off water and Carbon Dioxide.

But what is wax?
Maybe the problem is that we don't know what wax actually is.

Students can't believe wax burns because they have no idea that wax is a hydrocarbon. Students seem to need to know that wax is C and H to bolster their belief in burning.

So, I need to give them another half-page test to find out what they are thinking. I need to be really sure that they don't know. Then I can address their misconceptions.

What is wax made of?

In response to the half-page test question, the majority of students just said, "I don't know." A few tried to imagine what wax could actually be.

Student ideas: What Is Wax Made Of?

I'm not sure what wax is made of but I think it is made of maybe some kind of oil and hardener. Maybe from bees?

Something not burning but can melt

Some wax is made by bees but I don't know what this wax is made from.

I am not sure but it's created with heat to make it a solid form so it can not be burned as easily.

Hardened burnable liquid (not water).

Is wax made of something like clay? (They both harden).

Maybe it could be from some sort of animal part.

Some type of plastic?

I don't think I know what wax is made of.

I have no idea. (The majority of students said this.)

I think wax is made of atoms.

Maybe from bee hives?

Wax is made possibly of some kind of fuel.

I don't actually know but I know a long time ago candles were made out of lard which is sheep fat.

I think that wax is grown from a plant.

You heated and shaped it into what ever shape you wanted it to be.

Wax is made out of chemicals.

Wax is made out of oils.

Candles Are Dead Dinosaurs

Ask a kindergartner where milk comes from.
They will say it comes from the store.
We can smile in all our wisdom that they don't understand – yet.

Ok then. Where does wax come from?
What is wax really?
Our candles are not bees wax. They are paraffin wax.
Where does our wax come from.
Students are ready to listen. They want to know.
They have been wrong often enough and now they want to be right.

In a sense, candles are dead dinosaurs.
Very dead.
So dead they lost their Oxygen.
They are entirely Carbon and Hydrogen.
No longer carbohydrate.
They have become hydrocarbons.
From a long time ago.
The time of the dinosaurs. (Dinosaurs began 200 million years ago and died 65 million years ago)
Even before that time.
From 300 million years ago.
From the swamps of Pangaea.
Pond scum. Plants. Forests. Trees.
These turned to coal.
They turned to oil.
We pump it up and make it into car gasoline, asphalt.
And candle wax.

Wax comes from dead sea scum that settled to the seafloor and got covered with silt for about 300 million years. Wax is old (very old) sea scum. C and H sea scum. Dead algae. Dead sea plants. Dead carbohydrates. Dead Dinosaurs. Hydrocarbons.

"Hydrocarbon" and "carbohydrate" sound similar but they are different. Hydrocarbons consist of purely Carbon and Hydrogen, while carbohydrates also contain Oxygen. To be a carbohydrate, a compound must have twice as many Hydrogen atoms as Oxygen atoms, But they both have Carbon and Hydrogen.

Paraffin wax is a white or colourless soft solid derivable from **petroleum**, coal or oil shale, that consists of a mixture of hydrocarbon molecules containing between twenty and forty carbon atoms. It is solid at room temperature and begins to melt above approximately 37 °C (99 °F); its boiling point is >370 °C (698 °F).

https://en.wikipedia.org/wiki/Paraffin_wax

Well, if wax is C and H then maybe I can believe that wax actually burns. Wax has the components to make "worms with clothes on". When Oxygen does its job we say we have complete combustion. The C&H fuel combines with Oxygen (in the presence of heat) and becomes H20 and CO_2.

C&H + O_2 yields CO_2 + H20
We call that the equation for burning. It is simplified, of course, but that's the way it works.

Carbohydrates burn.
Hydrocarbons burn.
Coal (old plants turned into hydrocarbons) burns.

Now lets think about people.
We eat food and breathe in Oxygen.
We breathe out Carbon Dioxide and water.

C&H + O_2 yields CO_2 + H_2O
That's the same equation.
We say we "burn" our food – because we do.
The equation for living is the same as the equation for burning!
Wow!

Trees

We have been working with candles.
They burn.
They actually do burn.
Wax actually burns.

What about wood?
Here is a log.

Will it burn?
Will burning wood give off water?

Candles and alcohol burners do? Is wood like that?
Let's try it.

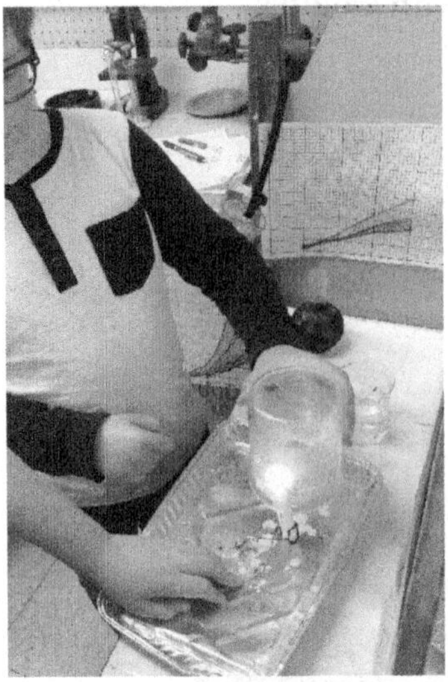

Yes, we can collect water over a wood fire.

Where does the water go?
Does it go into the air.

Is there water in the air right here?

We put water in a tuna can. We add some ice to cool it down.
We see water on the outside of the can.
Where did that water come from?

Many students believe the water somehow squeezes through the metal and comes from the water in the can.

How can I get them to understand that water is in the air all around us? Humidity. Dew point. These are important ideas. We could do experiments for a month to work out these ideas. I have to assess student enthusiasm. Is this the time to go into that? Maybe all I can do for now is just reference the idea. I ask what weather forecasters mean when they report humidity. Students are not sure.

This doesn't seem to be the right time to go into a detailed study of humidity. I decided to regale them with one last puzzlement.
We say fog lifts. Does it really lift?

No, not really.
Water stays right there but as the air warms, molecules move apart to allow the water to evaporate *right there.* Suddenly we can't see the morning fog but the water is *still right there in the air.* Invisible.

Even though there is not adequate time to explore humidity, I think our work with catching water above flaming candles and wood has given them the idea that there is water in the air. I am trying to establish that concept with the hope that some students will come to realize that trees can take that water out of the air.

Trees Are Made Of Air

What is in air?

Now I am stuck. Will I have to just tell them that our air is mostly Nitrogen? Fortunately, it turns out that most of them already know this. Air is mostly Nitrogen. What is Nitrogen?

Well, it is a gas. We don't need to know much more than that. At times like this, I sense that a few students may be interested to learn some details but they are not central to the argument we are developing. I usually try to find some interesting articles and pictures about the extraneous topic and tape them at eye level to shelves near where students enter the room. If they want to learn more, they can pause and read a bit. Some do. But most students just busily get their pencils and papers and sit down to start class.

The information I posted about Nitrogen included:

Nitrogen gas constitutes 78 percent of Earth's air.
It was first discovered by Scottish physician Daniel Rutherford in 1772. He removed Oxygen and Carbon Dioxide from air, demonstrating that the remaining nitrogen would not keep living organisms alive or let things burn. It was named after the Greek word *nitron*, for "native soda," and *genes* for "forming," Nitrogen is crucial to life and can be found in organic materials, foods, fertilizers, explosives and poisons. It is a colorless gas but it can easily be turned into liquid fertilizer. Bacteria in the soil change atmospheric nitrogen into ammonia, which plants need in order to grow. Animals eat the plants and get the nitrogen they need. Urine and feces return the nitrogen to the soil where bacteria change it back to gas which returns to the atmosphere.

Liquid nitrogen is also used in the manufacture of plastics, textiles, pesticides, dyes and in cleaning solutions.
Air also contains other gasses such as Oxygen, water and Carbon Dioxide.

Our class discussion returns to our central focus. I am trying to move students toward the equation for photosynthesis. And the equation for burning. I have to steer the ship to keep it on course. I have to make sure they are comfortable believing that air contains water and Carbon Dioxide. So I review.

We noticed water being given off by burning things.
On average, about 2% to 3% of the molecules in the air are water molecules.

Where does the water go? Does it go up?
Is there water in the air right here?
We put ice in a can. Dew formed on the outside – where did that water come from?

Is there Carbon Dioxide in the air?
We believe there is because we breathe out Carbon Dioxide.
Earth's atmospheric CO2 is currently 0.04% (about 400 parts per million)

Air is currently about 21% Oxygen.
Has it always been that way?
No, there wasn't much Oxygen in the atmosphere of our early earth. It seems that plants give off Oxygen and it is their activity that has created our current Oxygen level.

Hummm. Plants give off Oxygen. What do they take in?

(I'm leading them into trees. Into the equation for photosynthesis)

Jan Baptist van Helmont thought trees are made of water

Jan Baptist van Helmont was a Flemish chemist, physiologist, and physician. Born Jan. 12, 1580, died Dec. 30, 1644. He rejected the ideas of Aristotle that claimed the basic elements were; air, earth, fire and water and considered air and water the only true elements. He eventually identified Carbon Dioxide as a separate gas.

Van Helmont did many experiments but his most famous was about a willow tree. He placed a 5-pound tree in a pot containing 200 pounds of soil. He covered the pot so only rainwater from the air could enter the pot. After five years he weighed the tree (without leaves) and found it weighed 169 pounds. It gained about 32 pounds per year. The soil had lost only 2 ounces in weight. He concluded that the tree must be made of water.

His conclusion was wrong. A perfectly reasonable and logical conclusion … and perfectly wrong. But he had shown that the substance making up the tree had not come from the soil.

It is interesting no notice how we still respect scientists whose ideas we no longer agree with. We respect this man because he used the scientific method. His results may seem absurd. Comical. But he didn't have the framework of modern chemistry to work with. In fact, his conclusion was perfectly reasonable for the data he had. Interestingly, he went on to discover the gas Carbon Dioxide, the very substance overlooked in his famous experiment.

Well then, what are trees made of?
I need to know what my students believe now. Do they conclude as van Helmont did? I need them to commit their beliefs in writing and give them to me. Out of their control. They can't talk themselves out of it. They know I know and more importantly, they know they know. So, the question: **What are trees made of**?

Student ideas: What is a tree made of?

It starts as a seed, which grows from sun and water and maybe it has wood fibers in it, which expand and make a tree.

Wood if from a seed planted in soil, with water and sunlight.

A tree starts as a seed, then water it and give it sunlight and you have a tree.

You plant seeds and give it sun and water. It becomes a sapling, then a tree. People cut it to become a log.

A tree is made of quarks.

A tree is made of molecules and cells.

A tree is made of a seed and water and it grows.

Water, sun, seed.

A tree is made of many things I do not know but it is made of nutrients.

Seed plus sun plus water.

Are trees made of the same things wax is made of? Sun plus seed plus water. I think a tree is made out of carbohydrates.

I think a tree is made of cells and fibers and proteins and vitamins absorbed by leaves and roots.

A tree is made of Carbon, Hydrogen, Oxygen, stuff emitted from the big bang.

A tree I made of acorns that turn into other trees. Wood cells coming together.

I think a tree is made of bark and wood.

I think a tree is made of water, food, and sun.

The tree is made of an acorn, water, and rays of sunlight.

A seed and wood and water.

Trees Are Made Of Air

Wood. But I'm guessing that's not it so I'm not sure.

Water, seed, wood.

Wood, atoms, the big bang.

Matter, atoms, Oxygen?

Atoms

Atoms, molecules.

Wood, carbohydrates?

A tree comes from rich soil, water, sunshine, and an acorn or the seed.

A seed filled with fiber that, if exposed to air hardens into a crust-like substance called bark. Then the other fiber absorbs al the water and becomes wood.

A tree grows from a seed by getting water, soil for its roots, and an Oxygen environment. It is made of perhaps whatever comes in from its roots.

The seed absorbs small living matter that was pushed together to make fibers that make wood.

A tree is made of nature and atoms.

A tree is made by fiber absorbing Oxygen, water and vitamin D.

The seed absorbs water and grew into a tree. A tree is made of fiber.

A tree is actually made of water.

I think tree are made of fiber, water, sunlight, sugar and seed.

Seeds, fiber.

Fiber, molecules, atoms.

Water, sunlight, seeds, and bark.

Trees are made of fiber, but with the soil and water they're eating.

Trees Are Made Of Air

Trees Are Made Of Air

Someone in the class notices the equations that have been posted high on the wall in front of them all year.

$$CO2 + H2O \xrightarrow{\text{sunlight}} C\&H + O2$$

$$C\&H + O2 \text{ ----- } CO2 + H2O$$

What can these mean? Students are amazed. They have been looking at these statements all year but they had attached no meaning to them. They begin to explain to each other what the statements are saying. They are shorthand for everything we have been working on in this unit. Carbon Dioxide and water make trees. Trees are C&H and they give off Oxygen as a waste product that they don't need. They have to be out in the sun because sunlight is the glue that holds the H's onto the C's.

The second equation tells us about burning. And about eating food. We burn candles and wood. You have to have Oxygen in the air to burn things. We eat food. We have to breathe in Oxygen to "burn" our food.

The equations are opposite each other. The reactions go around and around. What is the point? Why bother? You just return to where you started.

Ah, but you don't. If we go out in the sun, we can starve to death. Sun is shining on our outside but it is not giving us our fuel. Our energy. But eating "food" puts that sunlight energy inside of us. When we "rip clothes off carbohydrate worms" we release the sunlight "glue" that was holding the molecule together. So really, we get energy from the sun by eating it!

We would not stay alive without plants. All animals would die without plants. When we eat hamburger, we are just eating meat that ate plants that packaged sunlight. All green plants do that. Trees do that. And they get their raw materials from the air, mostly. There are a few trace minerals needed from the soil, like vitamins. But trees are mostly made from air.

Now I tell my students that they are more knowledgeable than Harvard graduates. I tell them that most adults in their world do not know that trees are made of air. Carbon Dioxide from the air. Water from water vapor and rain – from the air. They are excited to learn that we will watch a YouTube video tomorrow showing people graduating from Harvard who say silly things like we said three weeks ago. My students feel very grown up and wise. They watch the videos with glee! They understand and Harvard graduates don't! Amazing!

The Annenberg YouTube video explains:
We have a major dilemma in our country. Despite twelve years of presenting school lessons that try to build children's understanding of the world through environmental inquiry, students still graduate from college not knowing much more than van Helmont did in 1600.

How can that be? The Annenberg foundation interviewed students graduating from Harvard, having just completed one of the best educations in the country and every interviewee failed to understand. They made a YouTube video showing the issue. They called the student's thinking their "*private universe*".

Private Universe Photos#106BE7F

Other YouTube videos show the same concern.

Where Do Trees Get Thei#106C0E0
This YouTube video says trees are 95% Carbon Dioxide.

Trees Are Made Of Air

Richard Feynman YouTube
https://www.youtube.com/watch?v=N1pIYI5JQLE

Feynman Fire FUN TO IMGINE YouTube
https://www.youtube.com/watch?v=ITpDrdtGAmo

Feynman explains that trees actually come from the air.

Light and energy from the sun bind Hydrogen to Carbon making carbohydrates in the process commonly known as photosynthesis.

The reverse happens when we burn those carbohydrates and hydrocarbons. We get back the energy of the sun.

$$CO2 + H2O \overset{\text{sunlight}}{\text{-----}} C\&H + O2$$

$$C\&H + O2 \text{ ----- } CO2 + H2O$$

Thus we come to the conclusion that ***trees are made of air.*** Students do not know how that happens. I tell them when they study cells next year they will have an opportunity to work with microscopes. Many of them will have the opportunity to put nail polish and tape on a leaf and make a slide of stomata.

I put some YouTube information and some pictures on the eye-level shelf. Maybe a few students will become interested. Some have microscopes of their own and are looking for interesting projects. Maybe they will discover stomata – those little holes in the underside of a green leaf.

The guard cells around the stomata open and close according to environmental conditions allowing them to take in Carbon Dioxide. Water vapor may move in and out of the open stomata and, of course, a lot of water from the air falls as rain and is absorbed by plant roots.

Stomata

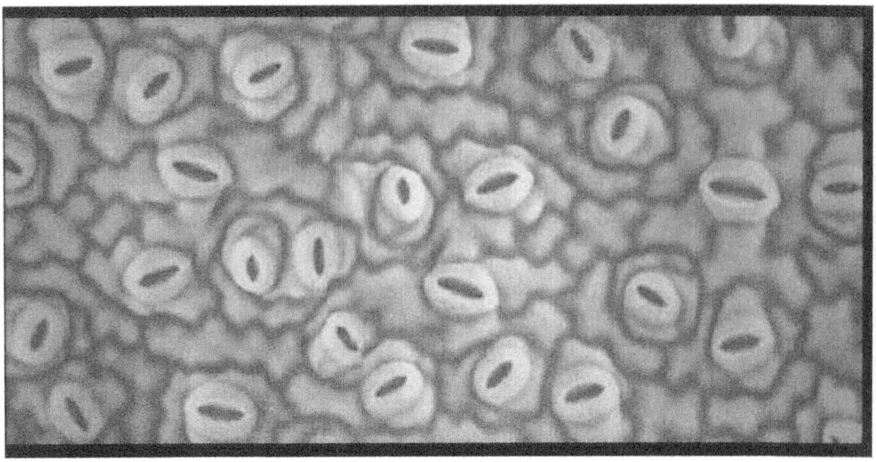

https://www.emaze.com/@ALWWQLOL/Science:-Photosynthesis

Here are some directions about how to make slides at home.

Stomata Printing: Microscope Investigation
https://www.calacademy.org/educators/lesson-plans/stomata-printing-microscope-investigation

Physical Change and Chemical Change

Enough talking. Students love being more knowledgeable than Harvard graduates, but they are still wiggly little kids. They are anxious to get back to doing labs.

There is one last big concept to be derived from all of this work. Surprisingly, standardized tests for this age child expect them to be able to spew back the terms *chemical change* and *physical change* even though they may have no grasp of what chemistry is about.

We have done enough work with burning and relating it to equations that my students may be able to grasp a beginning sense of it.

I try to explain. Burning and photosynthesis recombine atoms into entirely different appearing substances. We call that *chemical change.* Wax does not look like soot. Wood does not look like ashes. The Carbon has different "clothes" on.

When wax melts and drips into our pan, it is still wax. When we saw wood, we get sawdust. We call that *physical change*. The material is smaller but the atoms are arranged the same way.

Lets make a museum to show our understanding. Each student gets an 11 x 17 foil tray and inlays it with paper. Dividing it into two columns we title the sections *physical change* and *chemical change.*

Now find a material. Divide it into two pieces. Make a physical change in one section by breaking it into pieces. Tape that to the *physical change* column. Call Doc "O", "Ready to light!" Now singe the other half of the sample and douse it in the fire extinguisher beaker. Then tape it to the *chemical change* side of the museum paper.

Draw a line and add another material. How many materials can you find where you can cause both a *physical change* and a *chemical change*?

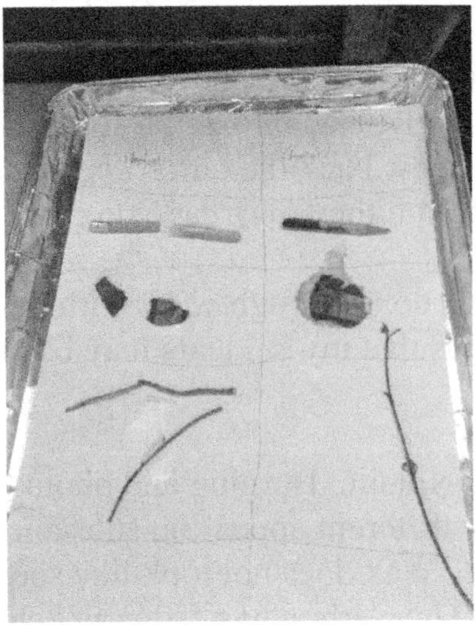

We Eat Sunlight

Time to turn in our units. We need to collect all our notes taken over the past three weeks. What shall we title the unit?

Maybe “Trees Are Made Of Air”.

Maybe “We Eat Sunlight”.

Maybe the two shorthand equations:

$$CO_2 + H_2O \xrightarrow{\text{sunlight}} C\&H + O_2$$

$$C\&H + O_2 \longrightarrow CO_2 + H_2O$$

The wonderful thing about equations is that they are our friends. They are not some fearsome, difficult thing to learn someday. They are a short way to explain or remind us of a month-long unit of study.

And just one more thing someone noticed:

$$CO_2 + H_2O \xrightarrow{\text{sunlight}} C\&H + O_2.$$

Takes in energy

$$C\&H + O_2 \longrightarrow CO_2 + H_2O$$

Gives off energy

The best and brightest students love fancy words.
Exothermic and *Endothermic* are fancy words.
They are opposites. One takes in energy, the other gives it off.

The cycle of energy goes round and round.
From trees to us and back to trees.

We are heaters.
And trees are refrigerators.

The gift of a concept simplified.
A simple enough overview to provide a framework for a lifetime.
With ample divergent pathways for future studies.
With opportunity for more detailed understanding as the student matures.

But a simple motto to frame that understanding.
To help me remember.
"Trees Are Made Of Air."

What did I mean by that?

Oh yes, trees are made of Carbon Dioxide.
And food is packaged sunlight.

About the Author

Meredith Olson Ph.D.

Dr. Meredith Olson, known affectionately as Doc "O" to her students, has taught elementary, middle school and high school math and science in Seattle for nearly 60 years. Her primary goal is in improvement of pre-college engineering education. By going to lab to work on contraptions every day, her students come to understand properties of the mechanical world.

"It has been a long and interesting trip. Studying some metallurgy in grad school. Evening classes. After a full day of high school teaching. Consulting for JPL as the Mars Pathfinder Educator. Weekends. Working in the summer with UNESCO in Zimbabwe, Kenya, and Uganda. Teaching dozens of weekend and week-long summer teacher workshops in South Carolina and Montana. Being a consultant and curriculum designer for Health and Physiology education in Washington, Oregon, Idaho, Montana, and Alaska. Being a summer adjunct University instructor for more than 20 years in Seattle, Idaho and Montana. Teaching teachers. Teaching students every day, every year for 59 years. Observing how learning happens. Becoming aware when real learning isn't happening. When it is just "show." When it is just teacher–pleasing to get a grade. To get a credit. To get a university degree."

See Dr. Olson's open letter outlining her philosophy of lesson design, available on the JPL website - Exploring Preface pp 11-13
http://mars.jpl.nasa.gov/education/modules/GS/GS07-19_preface.pdf

Dr. Olson believes that children must construct their own understanding from active design and assemblage of contraptions. By testing, failing, remodeling, and trying again, we come to see the structure when we look. By carefully examining materials we have, we may perceive how to use them in new and unexpected ways. Children begin to understand the engineering process. Besides, it is fun!

Made in the USA
Monee, IL
04 November 2026

67379928R00036